God's Lovable Leaves
...and Pine Trees!

Photosynthe-
what?

-sis!

Mary Ann Winslow

God's Lovable Leaves...and Pine Trees!

Book 9 of the God's Cool Creation Book Series

Copyright© Mary Ann Winslow 2023

ISBN: 9798218298548

Mary Ann Winslow, Author and Artist
Kent Winslow, Contributor

Cool Creation Press
Prescott, Arizona
www.godscoolcreation.com
Instagram.com/godscoolcreation
YouTube.com@coolcreationpress

To all the photosynthesis-challenged kids out there! Hope this helps! Remember, Jesus loves you!

Living things need to eat - that includes leaves! Our breathing helps to feed them! And they give us oxygen needed for us to breathe - thanks to leaves and the algae in the ocean!

How did God make all this happen?
PHOTOSYNTHESIS (foh toh SIN thuh sis)!

Inside the leaf is water. How does the water (H_2O - two hydrogens and one oxygen) get there?

Well, there are tiny tubes called xylems that carry water from the roots of the plant to the leaf. When the Sun hits the leaf, it splits apart that H_2O.

The leaf doesn't need the oxygen, or O, so it pumps it out from tiny holes on the underside of the leaf called stomata (STOH muh tuh). And we then use it to breathe! So that's how we get our oxygen! But how does the plant get fed by us?

It's all about carbon dioxide (CO_2) that we breathe out! It goes into the underside of the leaf in those stomata holes - the same place where the oxygen left - and it combines with the two hydrogens that were left from the water split by the Sun to make sugar for the plant! It's a little more involved, so let's get into it.

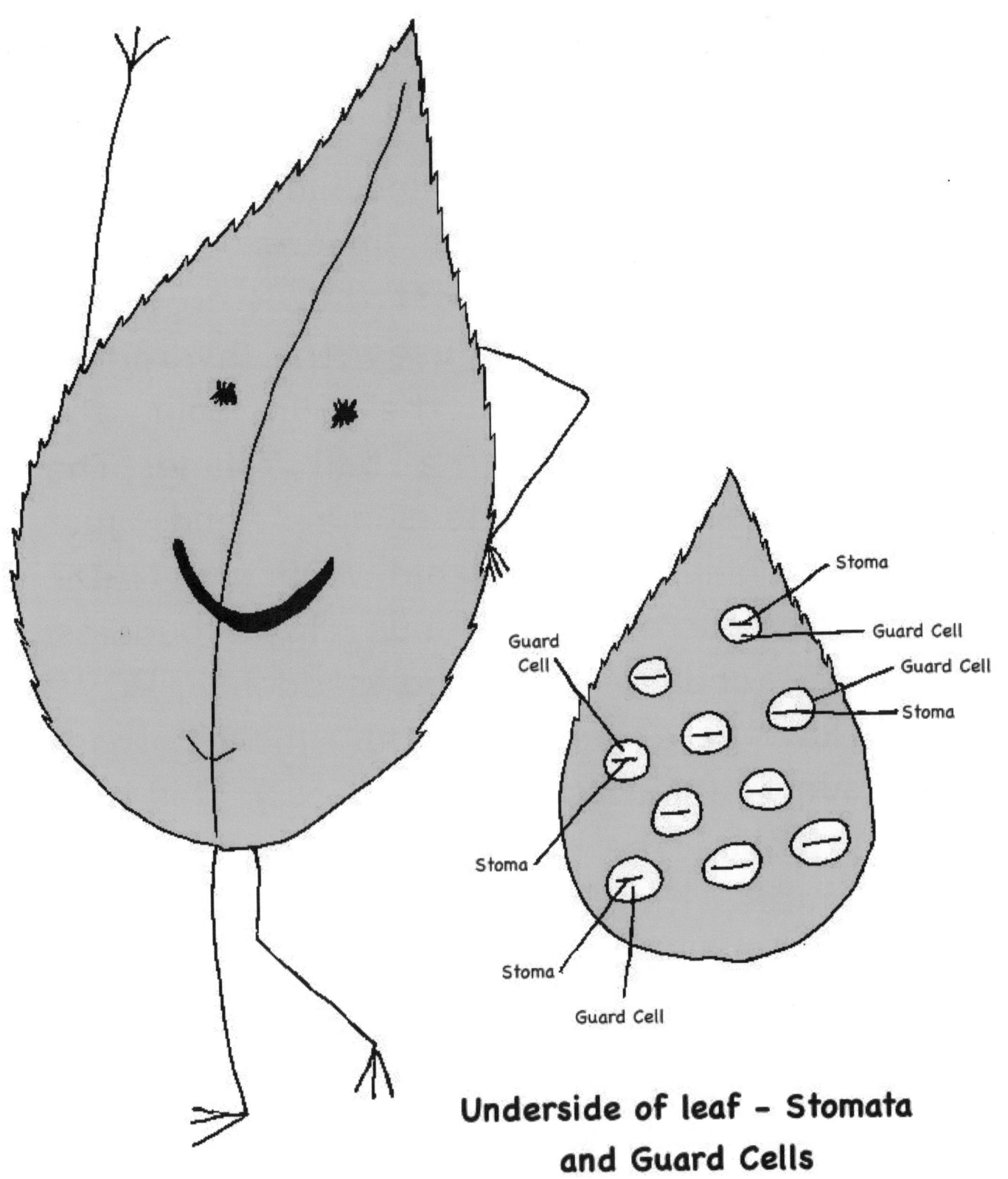

Underside of leaf – Stomata and Guard Cells

There's a tiny machine inside of that leaf that makes all this happen - a chloroplast (KLOR oh plast). This tiny machine uses the hydrogens left behind when the Sun split the water, and gets the ball rolling. The hydrogens attach to some friends just sitting in the thylakoid, ADP and NADP. They turn into real busy little buddies, ATP and NADPH. And suddenly, POW! Now there's power to start making those sugars (glucose - GLEW kohss). This part is called the light reaction because, of course, the Sun, God's light for Earth, starts it all.

Chloroplast

Now our little friends, ATP and NADPH, swim to another area, the fluid-filled space in the chloroplast called the stroma (STROH muh). They use our carbon dioxide and get busy making sugar! Thanks to these little friends, our plants can have a feast!

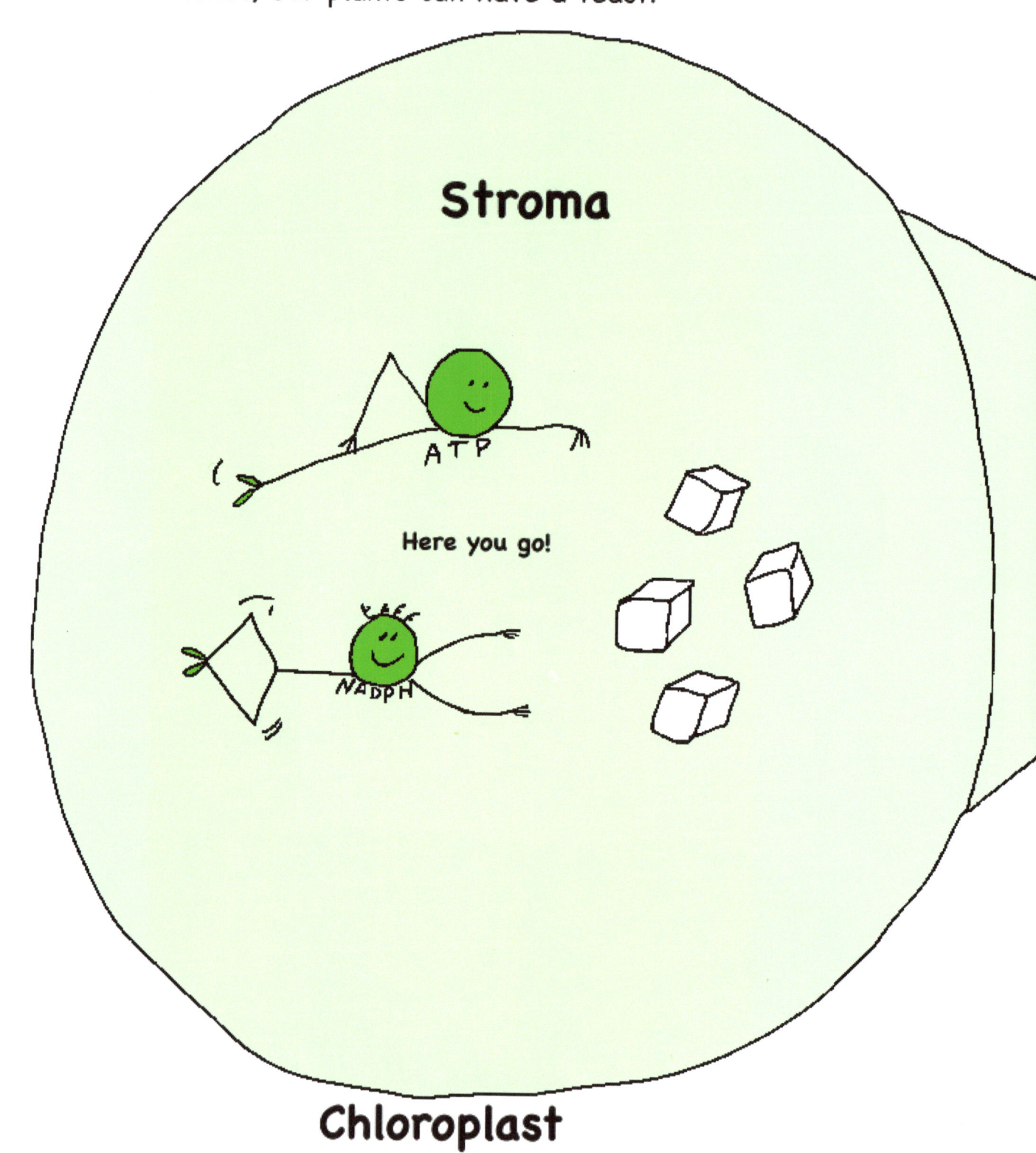

Stroma

ATP

Here you go!

NADPH

Chloroplast

This process is called the light independent reaction because it doesn't need light to make the sweets! From there, the sugar (glucose) is changed a bit and travels to stems, flowers and other parts of the plant via a little tube called the phloem (FLOW um)! God's amazing!

Uh-oh! How do plants like cacti in the desert, where it's really hot and there's not much water, start this process? Well, first of all, plants have to keep those stomata closed to save their water. There are special little cells called guard cells that make it happen. Guard cells open and close the stomata and guard what comes in and out. But wait! If they don't open, they don't get carbon dioxide! And we don't get oxygen! Not to worry...

Photosynthesis occurs on the saguaro skin!

Yes! And its thick, waxy skin reduces water evaporation!

God took care of this! The stomata open at NIGHT to take in our carbon dioxide.

Great, can we go home now so I can eat, too?

In the morning when stomata close because of the desert heat, another little machine uses the carbon dioxide stored from the night before to make sugars. Isn't it amazing how the Lord designed the plants so perfectly to survive the scorching heat?

How do pine trees undergo photosynthesis? Well, pine needles are leaves, and like other leaves, have stomata. But they have a thick coating so that they don't lose much water. And since the needles aren't as big as leaves, they can't take in as much light, anyway. They keep those needles all year round (for two years!), by the way, which is why they're called evergreens!

The fastest growing pine is the Eastern white pine. It grows two feet a year! The white pine needle tea is used to treat coughs and congestion, and helps with sore throats and kidney problems!

The longleaf pine, native to the southeastern United States, wins for the longest pine needles. They grow a foot and a half long! Its buddy, the loblolly, which also hails from the southeast, is famous for space travel! Its seeds were taken to the moon on Apollo 14. These "moon seeds" were then planted at the White House!

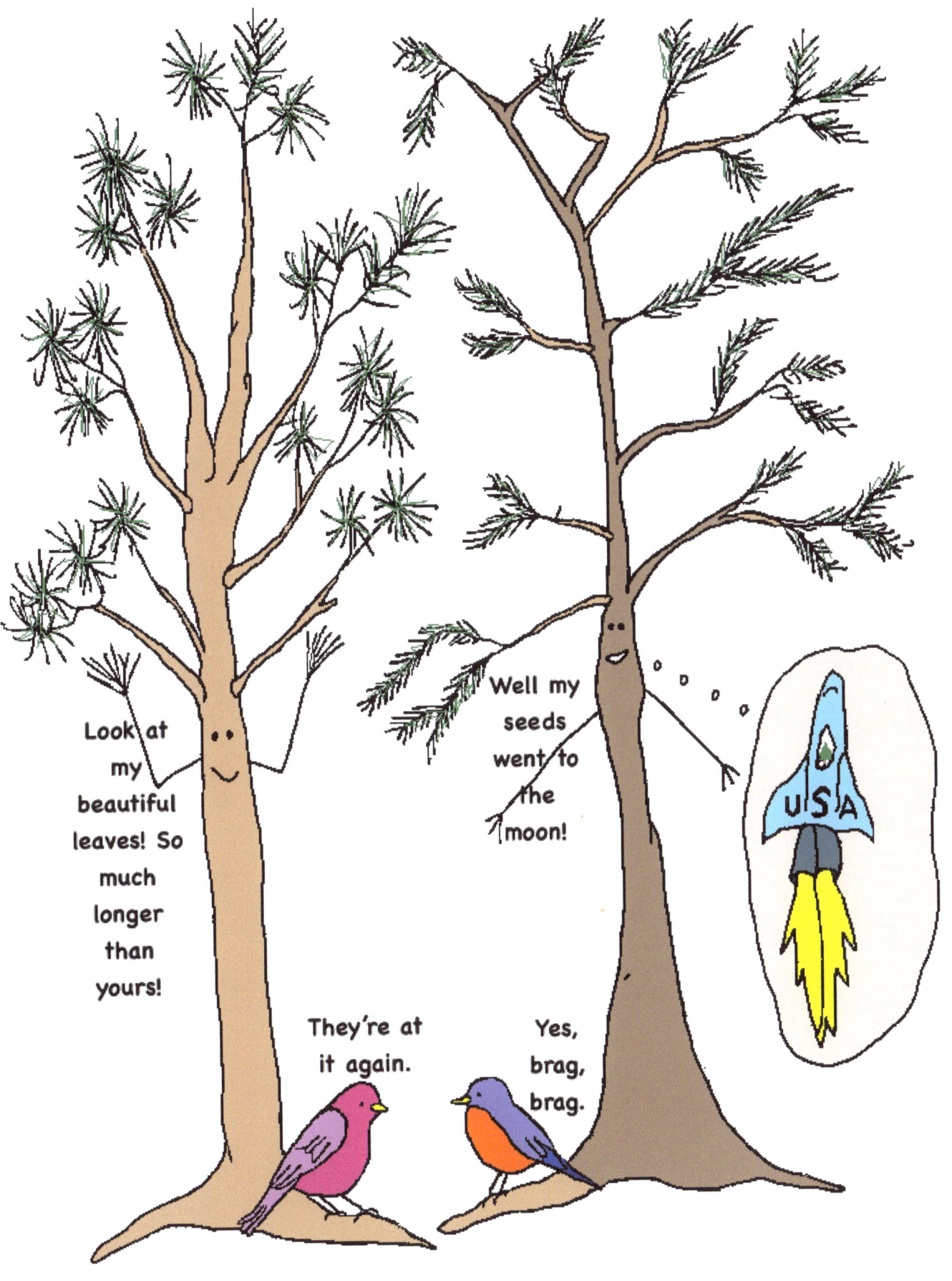

The loblolly, the tallest of the southern pines, grows to about 100 feet tall. That's the length of three school buses! The ponderosa pine of the western United States stands over 200 feet tall. That's the length of 20 African elephants! The sequoias and the redwoods of California, however, are the champions, reaching over 300 feet tall! That's the length of a football field!

The age of trees is measured by their rings. In wet years, the rings are wide and in dry years, the rings are narrow. The loblolly, for example, can live over 200 years! The ponderosa pine can live over 500 years! The sequoias and redwoods can live over 3,000 years! But the granddaddy and the grandmommy of them all is the...

Great Basin bristlecone pine! It can live over 5,000 years! That means that the same bristlecones, sequoias, and redwoods that exist today were alive when Jesus walked the Earth! Bristlecones live in harsh conditions. Temperatures drop well below freezing, the growing season is short, and they grow very slowly. This makes their wood very dense, so insects and erosion don't affect them. The harsh conditions and poor soil prevent other plants from growing there, so forest fires are rare. Few plants and trees around it mean that the bristlecones don't have to compete for water and nutrients – all the more reason for their longevity!

Only God could have designed how plants and trees eat, we breathe, and how He put so many species, or kinds, of trees (70,000!) on Earth for us to enjoy!